BLUEBELL STEAM
~ in Retrospect

1
Sir Archibald Sinclair climbs to Semley with an Ilfracombe
—Waterloo express on 8th June, 1964.

D. Cross

2
Southern Steam Personified! No 34023, bearing the name ▷
Blackmoor Vale, which it carried for the first few years of
its career, heads a West-of-England express near Brookwood
in the late nineteen-forties.

R. W. Beaton

Introduction

When the Bluebell Railway opened for business on 7th August, 1960, the operation of a standard-gauge passenger-carrying railway by volunteers was a venture into the unknown. Pessimists abounded, but they had not reckoned with the persistence and determination of the Line's founders, nor had they realised how popular the Railway would become with the public, a popularity which grew steadily as steam on the British Railway's system slowly disappeared. In 1968 when the Bluebell Railway purchased the freehold of the land, there could be no doubt that the Bluebell's future was secure.

To-day preserved steam railways abound throughout the country, and are now an integral part of the leisure industry. Some of these more recently established schemes can boast track mileages considerably in excess of the Bluebell's modest five mile stretch, and many are much larger operationally. Despite this the Bluebell Line continues to flourish with many thousands of visitors each year. What gives the line its special appeal?

By virtue of being so early in the field of standard gauge preservation the Railway was able to acquire a selection of vintage locomotives and coaches of considerable historic importance. Had it not been for the intervention of the BRPS in the early 1960s such elegant machines as the L&SWR Adams Radial Tank locomotive, and LB&SCR 0-6-2T locomotive No. 473 *Birch Grove* would have met an ignominious end in the breaker's yard. There is little doubt that locomotives such as these, now supplemented by more recent arrivals such as the SE&CR H class 0-4-4T and C class 0-6-0, contribute enormously to the Railway's delightful Edwardian atmosphere. So perhaps, it is the antique appearance of so many of the Line's veteran locomotives that gives the Bluebell Railway its special charm.

This album, produced to commemorate twenty years of Bluebell steam operation, looks back at the history of this unique collection of steam power. Nearly one hundred railway photographers have been contacted in the effort to trace suitable material, and many hitherto unpublished photographs have been submitted. Choosing the photographs has been an enormously difficult, and at times an extremely frustrating task. Whereas some locomotives, particularly the Bulleid Pacifics and Standard Class 4, appear to have been widely photographed during their main-line career, others have eluded the camera.

Throughout we have endeavoured to achieve a balance in the range of photographs selected, although those with a preference for action photographs will detect a bias in their favour. Some of the older photographs are of particular historic interest and have been included, although their quality may not be to the standard we would have wished.

The photographs are arranged in chronological order within each section, starting with No. 55, *Stepney*, the Bluebell's first locomotive, and ending with No. 34059, *Sir Archibald Sinclair*, which is, at the time of writing, the latest locomotive to be delivered.

Just before going to press it is learnt that another locomotive Southern Railway U class No. 31638 has been purchased privately from Barry scrapyard for restoration and use on the Line, and that a fund has been launched to save a BR 'Standard' Class 2, also from Barry. Perhaps these locomotives will be the first candidates for "Bluebell Steam In Retrospect - Part Two"?

Michael S. Welch
Worthing,
West Sussex.
April 1980

SBN 903899 03 5

3
The Bluebell Railway's Adams Radial locomotive is widely considered to be the most attractive on the line. It spent nearly thirty years of its life at Shepherdswell, on the East Kent Railway, before being purchased by the Southern Railway in 1946. The locomotive's elegant proportions can be clearly seen in this portrait taken at Shepherdswell in March, 1937.

J. M. Jarvis

Acknowledgement

We would like to express our appreciation to all those who have assisted during the preparation of this book.

Our thanks are offered to the following members of the BRPS who have helped in a variety of ways: Graham Burtenshaw, Terry Cole, Roger Cruse, Mike Esau, Bernard Holden, Mike Mason, Klaus Marx, George Nickson, Ian Osbourne, Roger Price, Chris Pyle, Pete Reid, David Wigley and Martin Wilkins.

We are also most grateful to David J. Fakes who supplied a great deal of the historical information, and to Jennifer Sanders and Derek Hedger who typed the manuscript. The excellent printing of Mr. Derek Mercer of Messrs Neville Dexter, Streatham, London, has done much to enhance many of the photographs.

Finally, we would like to record our gratitude to the many photographers who have so kindly allowed their work to be published, and also to those who submitted photographs which were not included in the final selection. It is a pity there was not space for them all!

M.J. Allen
M.C. Frackiewicz
M.S.W.

Published by
Bluebell Railway Preservation Society
Sheffield Park
East Sussex

Design by Oxford Publishing Co.
Typeset by Katerprint Co. Ltd., Oxford
Litho Reproduction by Knight Publishing Ltd., Oxford
Printed by B. H. Blackwell in the City of Oxford

Contents

4

Stepney, made famous as 'Stepney the Bluebell engine' in the Reverend Awdry's series of children's books, is probably the best known of all Bluebell engines. It entered service at New Cross depot as No 55 on 21st December, 1875 where it was put to work on suburban duties. There were fourteen 'Terriers', or 'Rooters' as they were commonly known, based at the depot and they shared a roundhouse specially built to house them. This picture, the oldest known photograph of *Stepney*, was taken on 20th September, 1902 at New Cross shed.

LCGB Ken Nunn collection

LB & SCR 0-6-0T No 55 'STEPNEY'

5

These suburban duties outgrew the 'Terriers' and at the turn of the century No 655—she was re-numbered in 1901 —migrated south to Brighton; it was photographed at the erstwhile London Road carriage sidings in about 1904.

M. P. Bennett

6
At Brighton Station, circa 1906.
M. P. Bennett

7
The next thirty years were eventful ones in the life of *Stepney*. It returned to South London for a while, but in 1920 found itself on loan to the Woolmer Instructional Military Railway, forerunner of the Longmoor Military Railway. In 1925, No 655 was withdrawn from traffic as surplus to requirements and stored in Preston Park paint shop. It was returned to traffic in January, 1927, following a repaint in SR green livery, and was sent to Fratton shed for use on the Lee-on-Solent branch. By the mid-thirties however its duties from Fratton were mainly on the Hayling Island branch, and in this picture it is depicted sprinting along between Fratton and Havant on the 8.47 a.m. Sunday morning Portsmouth Town–Havant train in August 1936. This train was the last regular passenger-carrying working for a 'Terrier' on the main line.
Dr. Ian C. Allen

8
No 2655 at Fratton MPD on 19th June, 1937.

J. G. Dewing

9
After spending the war years based at Rolvenden on hire to the K & ESR, *Stepney* returned to Fratton for a few years for a further spell on the Hayling Island branch. Here it powers a Havant bound working on 1st September, 1951.

B. E. Morrison

10
By 1953, No 32655 was back on the K & ESR, and is depicted at Bodiam on 12th August, 1953, while working the 12.20 p.m. Robertsbridge—Tenterden mixed train.

G. F. Bannister

11
Later the same day *Stepney* stands outside Rolvenden shed.

G. F. Bannister

12
On 2nd January, 1954, the K & ESR closed to passenger traffic, and here No 32655 is seen curving away from the main line at Robertsbridge with one of the last passenger workings.

W. M. J. Jackson

13
Stepney's final six years of BR service were spent at Brighton, and during this time it busied itself on a variety of local duties. On 22nd March, 1957 No 32655 was photographed shunting the West Quay line at Newhaven.

W. M. J. Jackson

14
Stepney at Sheffield Park on 8th October, 1960. Note the engine shed under construction in the background.

D. W. Winkworth

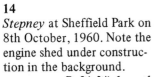

15
Now fully restored, *Stepney* was photographed at Sheffield Park on 26th May, 1963. The first vehicle in the train is the LNWR Observation Car; the other coaches are a rake of BR Standard and Bulleid vehicles which had been loaned to the Bluebell while the Metropolitan coaches were away in London in connection with the Centenary of the Metropolitan Line.

R. C. Riley

16

No 323 was one of a batch of eight P class tank locomotives constructed in 1909–10 at the South Eastern and Chatham Railway Company's Ashford Works. They were intended for use on push-pull trains and lightly-laid branch lines, and were similar in many respects to the LB & SCR 'Terriers' on which they were modelled. Unlike their Brighton counterparts, the P class engines were not wholly successful, and eventually found themselves relegated to light shunting and shed pilot duties. In this picture No A323 stands at Dover shed in 1929 having been transferred there in 1926 following the electrification of the Otford–Sevenoaks line.

J. A. G. H. Coltas "Locofotos" collection

SE & CR 0-6-0T No 323

17
No 1323 in later Southern livery.
Lens of Sutton

18
No 31323 in final BR condition; this picture was taken at Dover on 1st June, 1958.

P. H. Groom

19
When No 31323 came to the Bluebell in June 1960 there were misgivings amongst many members who would have preferred another 'Terrier'. BR however had none of the latter engines available, and the Bluebell were obliged to consider a P tank instead; it was with some reluctance therefore that the Bluebell accepted No 31323. All doubts were dispelled when it proved a very capable performer, being consistently free-steaming, reliable and economical. It is generally accepted that for the first ten years of the line's existence the Ps were the mainstay of Bluebell operation.
No 323 passes Polegate en route from Ashford Works to Sheffield Park on 26th June, 1960.

S. C. Nash

20
The best time to find one of the P class locomotives in passenger use is during the Winter when train loads are usually lighter. Here No 323's exhaust hangs in the air as it rounds the bend at Freshfield with the BR Standard set of coaches on a bitterly cold New Year's Day 1979. Since its arrival at Sheffield Park No 323 has carried the name *Bluebell*.

M. J. Esau

21

No 3 *Baxter* was built in 1877 by Fletcher Jennings of Whitehaven, Cumberland, for shunting duties at the Dorking Greystone Lime Company's Betchworth works. It is now the sole remaining standard gauge locomotive produced by that builder. Like *Stamford*, No 3 was built specifically for use at one location and *Baxter* worked at Betchworth from its introduction until 1959 when the quarries closed. The engine ran un-numbered until the late nineteen-twenties when it became No 3 and was named *Captain Baxter*, after the Chairman of the company. In 1947 it was re-christened simply *Baxter*. It was placed on loan to the Bluebell Railway in 1960 by the late Major E. W. Taylerson of the Dorking Greystone Lime Company, and subsequently Major Taylerson's executors have kindly agreed to this interesting little engine remaining on Bluebell metals. This photograph is thought to have been taken shortly after the locomotive's arrival at Betchworth.

Courtesy F. Webb

FLETCHER JENNINGS & CO 0-4-0T No 3 'BAXTER'

22

On the last Saturday of the 1960 season *Baxter* made its one—and so far only—sortie on passenger duty in connection with the visit of a well-known railway historian. Here it is seen awaiting departure from Sheffield Park watched by a group of amused onlookers.

D. Alexander

23
Unfortunately no pictures of any Bluebell P class locomotives in SECR lined green livery appear to have survived; this is No 27 in wartime grey livery. The location is probably Dover shed; the date is unknown. During the First World War No 27 was one of two P class locomotives sent across the Channel to work at Boulogne.

Rail Archive Stephenson

24
Two Bluebell P class locomotives for the price of one! This photograph was taken at Dover in 1929.

J. A. G. H. Coltas "Locofotos" collection

SE & CR 0-6-0T No 27

25
With its previous number and ownership visible beneath faded paintwork, No 31027 reposes on Eastleigh shed on 27th August, 1952.

J. Kent

26
No 27 passes Hassocks en route to
Sheffield Park on 18th March,
1961.

S. C. Nash

27
No 27 lays a smokescreen across
the Sussex countryside near
Freshfield Halt as it partners
Fenchurch on a northbound train
in May 1974.

D. T. Cobbe

28
No 488 probably has one of the most colourful histories of all Bluebell locomotives. It was built in 1885 by Neilson and Co. Ltd., Glasgow, for the London and South Western Railway who used the class on suburban services out of Waterloo. In the early years of the twentieth century the 0415 class had been superseded by more modern classes, and by 1903/4 most had been transferred elsewhere, although some lasted there until electrification. In September 1917, No 0488, as it became, was sold out of service to the Ministry of Munitions, who put it to work at Ridham Salvage Depot (Sittingbourne) where it is pictured in 1918. On the right is 0-4-2T *Skylark* which was also used by the Ministry of Munitions at this time.
R. Godman: Courtesy SKLR Ltd.

L & SWR 4-4-2T No 488

29
In April 1919 the locomotive was re-sold to the East Kent Railway for the sum of £900, and was allotted the No 5 in their series. It is seen here at Shepherdswell in June 1936, looking resplendent following a re-paint.

D. Middlemiss

30

Meanwhile, all the 0415's on the main system, apart from two survivors employed on the Lyme Regis branch, had been withdrawn. In 1946 both these machines required repair, and the SR approached the EKR regarding No 5. By this time fate had overtaken the ageing 4-4-2T and it had lain dis-used at Shepherdswell since 1939. Despite its rusty exterior the locomotive was found to be repairable and the SR offered the EKR £800 for the engine which was readily accepted. On March 15th, 1946, No 5 left Ashford for Eastleigh Works where it was given a thorough overhaul. It was officially taken into SR stock as No 3488 on 13th August, 1946, and was turned out in Southern livery for the first time. It is seen at Exmouth Junction in 1949, taking a break from Lyme Regis branch duties.

W. M. J. Jackson

31

For the remainder of its BR career No 30583, as it became under the BR numbering scheme, was confined to the Lyme Regis branch, and here it posed for its picture on 2nd September, 1952.

A. N. H. Glover

32

On 12th April, 1953, it worked a special train from Exeter to Exmouth via Sidmouth Junction, which was recorded entering Exeter Central at the completion of the tour.

E. D. Bruton

33
No 30583 was again engaged on special duty when on 28th June, 1953, it powered two specials down the Lyme Regis line in connection with the 25th anniversary of the RCTS. It is seen here leaving the terminus assisting ex-LBSCR A1X No 32662. The tours were booked to be powered by two 'Terriers', but one of these failed and No 30583 was a last minute substitute.

E. D. Bruton

34
The shadows are lengthening as No 30583 enters Lyme Regis with an afternoon train from Axminster on 29th August, 1956.

P. H. Groom

35

A delightfully clean No 30583 bathes in the sun at
Axminster between trains on 4th August, 1960.

P. H. Groom

36

The Autumn sun glints on the side of the train as No 30583
jogs along on a one coach Lyme Regis - bound working in
October 1960.

S. C. Crook

37

After withdrawal from service No 30583 lay at Eastleigh for several months before being despatched to the Bluebell. It is seen here passing Warblington Halt en route to Brighton and its new home on 9th July, 1961.

J. C. Haydon

38

No 488 worked regularly on the Bluebell from its arrival in July 1961 until 1967 when it became due for heavy repairs. Unfortunately the work required to restore the locomotive to working order was beyond the capacity of the Railway's then meagre facilities, and it lay stored out of use at Sheffield Park until October 1971 when it was removed to Swindon Works for an extensive overhaul. The locomotive was completely dismantled, the boiler being despatched to Crewe for attention. It is pictured here in Swindon Works on 20th December, 1972, well on the way to re-assembly and eventual return to the Bluebell.

R. C. H. Nash

39
Looking immaculate following a repaint No 488 worked a special train to mark its return to active use on 4th August, 1973. Here the special rounds the curve at Freshfield on the outward run.

D. T. Cobbe

40
No 488 starts the climb up Freshfield bank with the 3.55 p.m. departure from Sheffield Park on 18th July, 1976.

S. C. Nash

41

The 90xx class, commonly known as 'Dukedogs', were constructed by the Great Western Railway at Swindon Works in 1936. After the grouping the GWR found that the 32xx 'Duke' class was ideally suited for the services over the former Cambrian Railway, but by GWR standards the class was obsolete. At the same time the relatively modern 'Bulldog' 34xx class was redundant, having been superseded by the latest six-coupled classes. The 'Bulldogs' were too heavy for the Cambrian Lines, but by combining the 'Duke' boiler with the 'Bulldogs' frames a relatively modern lightweight 4-4-0 was obtained. The 'new' 32xx was created, and here seen at Swindon Works in 1938 is No 3217; this picture is thought to have been taken shortly after its construction. No 3217 was re-numbered 9017 in 1946 when the classes' number series was required for a new type.

P. F. Winding

GWR 4-4-0 No 3217

42

On special duty, No 9017, accompanied by sister engine No 9021, stands at Ruabon prior to departure for Porthmadog on a Festiniog Railway Society special. The pair had just taken over from No 3440 *City of Truro* which can be seen on the right of the picture. The date is 26th April, 1958.

P. F. Winding

43
Later the same day the 4-4-0s are seen departing from
Minfford. The Festiniog Railway Station is partially
hidden behind the trees.

P. F. Winding

44
No 9017 is pictured at Machynlleth MPD on 26th September,
1959. The two gentlemen are the shed master (on the left),
and an enthusiast who was involved in the acquisition of the
locomotive for preservation.

R. J. Blenkinsop

45
Summer 1960 was No 9017's last in BR service. On 27th
August, it powered a Butlins holiday special and was
photographed skirting the River Dovey at Glandyfi.

J. C. Haydon

46
No 9017 poses outside Machynlleth depot later the same day.
Strangely three Bluebell locomotives—the others are Nos
75027 and 80100—have been associated with the Cambrian
lines during their BR career.

J. C. Haydon

47

Dovey Junction on 30th September, 1960, and No 9017 waits to depart on one of her final BR workings—the 11.05 a.m. to Pwllheli. The following month she was withdrawn and went into store at Oswestry Works to await purchase by a private fund instituted for the purpose.

R. C. Riley

48

Eighteen months later No 9017 had been purchased—one of the first examples of a locomotive being saved by voluntary effort—and was despatched south to the Bluebell Railway. On arrival on the Southern Region it was found that the 'Dukedog' was facing the wrong way round, the Bluebell required to have her facing north and a trip was made to Brighton for turning purposes. Obviously the word about No 9017's visit had been passed down the 'grapevine' to at least one local enthusiast. Here the 'Dukedog' awaits departure from Brighton shed after turning.

E. Wilmshurst

49
On the Bluebell No 9017 has reverted to its
proper number, 3217, and bears the name
Earl of Berkeley, which was originally
allotted to the locomotive but never carried.
The 'Dukedog' is pictured near Freshfield
on a train bound for Sheffield Park.

M. J. Esau

50
In 1977 a group of enthusiasts repainted
No 3217—which was out of service for heavy
repairs—in BR livery as No 9017, in order to
simulate a Cambrian line freight train of the
1950s. Here No 9017 is seen standing at the
foot of Freshfield bank after being propelled
from Sheffield Park by another locomotive
which had been discreetly hidden around the
corner!

J. Goss

51
Reflection at Cromford wharf. No. 58850 shunts the
yard at Cromford on 25th April, 1953.

E. D. Bruton

L & NWR 0-6-0T No 2650

No 2650 first saw the light of day at the North London Railway's
Bow Works in 1880; it was designed for dock shunting and freight
work. When new it was No 76 in their series, but in 1891 it was
rebuilt and renumbered 116. In 1909 the NLR was taken over
by the LNWR and No 116 was renumbered 2650; in turn the
LNWR was absorbed by the LMSR and it became No 7505 in
1926. In 1931 the first of the dock tanks were moved away from
London to Derbyshire, to work on the Cromford and High Peak
line on which they were to perform so well. Three years later the
locomotive was again renumbered, becoming No 27505, and
eventually No 58850 under the BR regime in 1948.

52
A pair of North London tanks
provide the motive power for a
Gloucestershire Railway Society
special train on 21st May, 1955.
The scene shows Nos 58850 and
58856 waiting at Middleton Top
while the enthusiasts climb aboard
their train of open wagons.

D. H. Ballantyne

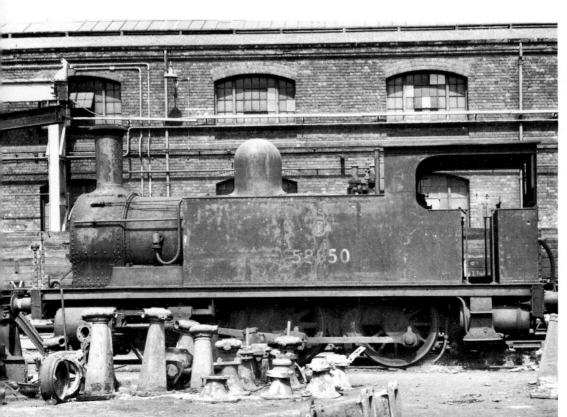

53
No 58850 appears doomed as it
sits at the back of Derby Works
on 18th June, 1961, a year after
withdrawal.

A. N. H. Glover

54
Fortunately No 58850 was rescued by the Bluebell Railway, and arrived at Sheffield Park on 28th March, 1962. In 1964 it assisted in the demolition of the line between Horsted Keynes and East Grinstead, and is seen here stabled at West Hoathly between track lifting duties on 4th October, 1964.
E. Wilmshurst

55
A closer view of No 2650 at West Hoathly at the end of the day's activity.
E. Wilmshurst

56
The Bluebell is always in demand for filming purposes and
in this view No 2650 is seen disguised as a ghost train for
the film 'I'll Never Forget What's-'is-Name' on
6th March, 1967.

M. J. Mason

57
No 473 was one of the 75 E4 class engines designed by R.J. Billinton for the LB & SCR. It was the eleventh locomotive to be constructed, and on completion in 1898 was sent to New Cross shed. The class was named after towns or villages within the parent Company's area, and No 473 was allotted the name *Birch Grove*, a country house, which, coincidentally, is situated close to Horsted Keynes. This picture of No 473 was taken at New Cross shed and shows the engine in original condition with the Stroudley type livery.

M. P. Bennett

58
No 473 in early Southern livery.

Lens of Sutton

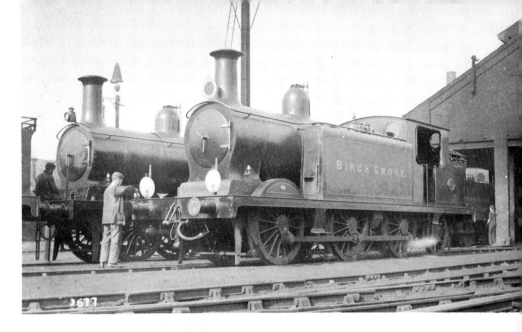

LB & SCR 0-6-2T No 473 'BIRCH GROVE'

59
In the days before BR had organised their locomotive numbering system some engines were simply given an 's' prefix to their pre-nationalisation number, although the 'British Railways' legend was applied to the tanksides. It is interesting to note that at this time there were no transfers available and the tanksides had to be hand-lettered. s2473 is seen at Bricklayers Arms on 30th May, 1950.

J. Kent

60
No 32473 in final BR condition poses at Feltham shed on
29th August, 1961.

P. H. Groom

61
The E4's last duties on BR were confined to working empty
stock trains between Clapham Junction carriage sidings and
Waterloo. Here No 32473 stands at Clapham Junction on
25th April, 1962.

J. Scrace

62

On 27th October, 1963, *Stepney* and *Birch Grove* powered a special train from Brighton to Sheffield Park to mark the closure of the Haywards Heath–Horsted Keynes line. Here No 473 is seen moving off Brighton shed's turntable after turning.

S. C. Nash

63

Later the same day the pair were photographed near Patcham. Note the magnificent rake of vintage coaches which formed the train. The first vehicle is a L & SWR 'Ironclad', and the rest of the vehicles are of SE & CR origin. At this time this set was still regularly used on the Lancing workman's train between Lancing Carriage Works and Brighton.

S. C. Nash

64

Birch Grove makes an attractive picture in the Spring sunshine as it heads towards Horsted Keynes. This photograph was taken between Freshfield and Holywell on 5th May, 1968.

J. G. Mallinson

65

No 473 ascends Freshfield bank on 26th September, 1970. The first vehicle in the train is an SE & CR ten compartment coach of 1922 vintage.

M. J. Esau

66

The Blue Circle is a most unusual machine – the Bluebell stock book describes it as a "mechanical curiosity"—and it is now one of the very few survivors of its type. It was constructed by Messrs. Aveling and Porter of Rochester, Kent, in 1926 for shunting the works of the local Holborough Cement Company, and in many ways is reminiscent of that firm's road locomotives and rollers. It spent its entire working life at the Holborough works and was in use until 1963, though latterly only as a stand-by locomotive. In 1964 the Associated Portland Cement Company, successors to the Holborough Cement Company, presented the locomotive to the Bluebell Railway. At this time it was appropriately bestowed with the name *The Blue Circle*, the Company's trade mark. It is seen here at Holborough works on 7th April, 1956.

C. Hogg

AVELING & PORTER 2-2-0WT 'THE BLUE CIRCLE'

67

There is no doubt that the novelty value of this machine is immense, and at the time of writing it is undergoing painstaking restoration by members of the London Area Group of the BRPS. Here it is seen attracting considerable attention at Sheffield Park in 1964.

K. D. Chown

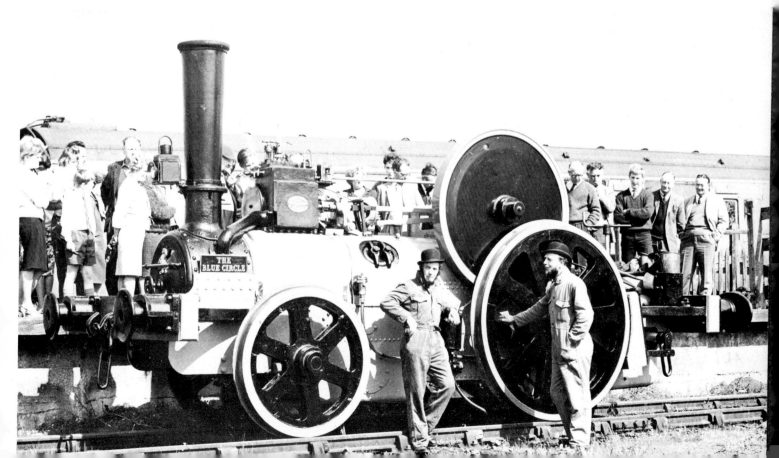

LB & SCR 0-6-0T No 72 'FENCHURCH'

68

No 72, *Fenchurch*, is the oldest locomotive on Bluebell metals. It was built at the Brighton Works of the LB & SCR and entered traffic on 7th September, 1872. This official Company photograph was taken outside the Brighton paint shop shortly after the locomotive entered service: note the three link couplings, and lack of Westinghouse equipment. The construction cost for *Fenchurch* was officially stated to be £1,800 18s 11d. The gentleman in the cab by the controls is thought to be William Stroudley, the locomotive's designer.

Courtesy: National Railway Museum
York

69

On entering traffic *Fenchurch* was sent to Battersea depot from where it worked suburban trains along the South London line. It shared these workings with five other 'Terriers', and the locomotives were an undoubted success on these duties, where they worked sets of close-coupled, light-weight high-capacity stock. So successful were these 'Terrier'- operated sets, that even up to the turn of the century the locomotives were often seen working trains rostered for larger motive power. A few years after its arrival at Battersea *Fenchurch* was fitted with a Westinghouse air pump and this picture, thought to have been taken in the late eighteen-seventies, shows No 72 posing in Battersea yard with this newly acquired apparatus in position.

Courtesy: National Railway Museum
York

70

As train loads increased these suburban duties outgrew the engines, and they began to move out of London to country depots for branch line work. Some were disposed of to contractors and other railways; No 72 was sold to the Newhaven Harbour Company in 1898. In 1913 *Fenchurch* was rebuilt to class A1X with a new boiler, and led an uneventful life at Newhaven where it became known as the "Tram Engine" due to the fact that the West Quay sidings there had originally been built as a tramway. This photograph shows *Fenchurch* in post-1922 Newhaven Harbour Company livery outside Newhaven Town engine shed.

Lens of Sutton

71

In 1926 the Southern Railway absorbed the Newhaven Company and in December of that year *Fenchurch* was despatched to Brighton Works for overhaul, later emerging as No B636—the lowest vacant number in the 'Terrier' series. In February 1927—still black, but now in Southern livery—the locomotive resumed its duties at Newhaven.

Courtesy: National Railway Museum
York

72

In 1935 a small supply of LB & SCR gilt numerals was discovered, and as No 2636 in the Southern Railway renumbering scheme, *Fenchurch* was fortunate enough to receive the benefits of this windfall. The numerals were retained on the locomotive until 1950. This picture was taken at Newhaven in 1948.

P. S. Leavens

73

In July 1952 *Fenchurch* made a return to passenger work, its first for over fifty years, and the following October was used on a shuttle service from Brighton to Kemp Town which operated in conjunction with an RCTS rail-tour from London. Here No 32636 enters Brighton hauling one of the Kemp Town specials on 5th October, 1952.

J. G. Dewing

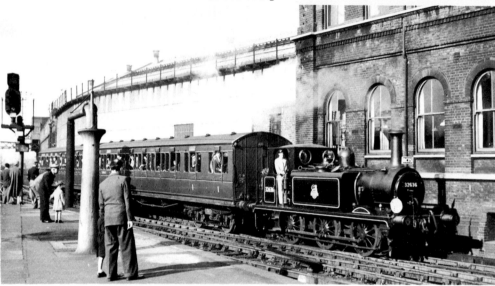

74

On 13th April, 1958, LB & SCR Atlantic No 32424, *Beachy Head*, made a commemorative final run from Victoria. *Fenchurch* was one of the locomotives waiting to receive the Atlantic on its arrival at Newhaven, the others being Standard Cl. 4 tank No 80154, the last locomotive built at Brighton Works, and 'Terrier' No 32640 (on the extreme right) the 1878 Paris gold medallist.

P. Winding

75

No 32636 acting as Works pilot at Brighton in the Summer of 1961.

G. R. Siviour

76

On 7th October, 1962, No 32636 was again used on a special train organised by the RCTS. It is seen climbing away from Falmer in the Autumn sunshine; the train engine is E6 0-6-2T No 32418.

E. Wilmshurst

77

The arrival of a pair of USA 0-6-0Ts at Lancing Works in June, 1963, coupled with the closure of the West Side lines at Newhaven two months later, deprived the 'Terriers' of most of their remaining work and No 32636 was despatched to Fratton to assist its sister engines on the Hayling Island service. On June 30th, however, No 32636 was pictured at Fratton receiving attention, presumably to a hot bearing sustained on its journey from Brighton.

D. Capewell

78

No 32636 simmers inside Fratton shed on 2nd November, 1963. In the background are Z 0-8-0T No 30952, and 'King Arthur' class 4-6-0 No 30777, *Sir Lamiel*. At this time the Z appears to have been set aside for preservation, but this did not materialise and the locomotive was subsequently broken up.

D. Capewell

79
The last train to Hayling Island,
3rd November, 1963.

J. G. Dewing

80
Another shot on the same day. The
coach immediately behind the
engine is a BR experimental glass-
fibre bodied vehicle.

T. Stephens

81
Fenchurch makes a splendid sight in the Winter sunlight as it climbs towards Horsted Keynes on 4th January, 1970.
J. G. Mallinson

82
In August 1975 *Fenchurch* represented the Bluebell at the celebrations to mark the one hundred and fiftieth anniversary of the Stockton and Darlington Railway. It stands on display at Shildon on 26th August, 1975.
E. Crawforth

83

No 75027 was one of 80 mixed-traffic Cl. 4 4-6-0s designed at Brighton Works to the specification of R. A. Riddles. No 75027 was built at Swindon Works in 1953. Early in its career it was allocated to the WR and saw service at Oxford depot for a time in the fifties. It is seen here near Knowle and Dorridge, on the GWR Leamington—Birmingham line, heading a down Morris Cowley fitted freight on 8th July, 1958.

M. Mensing

BR STANDARD 4-6-0 No 75027

84

In 1960 No 75027 was moved away from Oxford to Templecombe shed on the Somerset and Dorset line. The Cl. 4s' prime function on the S&D was the piloting of heavy holiday trains over the Mendips, from which duties they ousted the last of the class 2P 4-4-0s. Here one of the 2Ps stands in Templecombe shed yard as No 75027 and 'West Country' class 4-6-2 No 34039, *Boscastle*, pass on the 10.20 a.m. Liverpool (Lime Street)—Bournemouth West. This picture was taken on 1st July, 1961.

J. C. Haydon

85
No 75027 assists rebuilt SR Pacific No 34028, *Eddystone*, away
from Shepton Mallet with a Manchester—Bournemouth express
on 9th September, 1961.

Ivo Peters

86
During the Winter months No 75027 had to
content itself with service on stopping trains.
It stands at Bath Green Park on 9th March,
1962.

J. A. G. H. Coltas
"Locofotos" collection

87
No 75027 is seen again on pilot work this time surmounting Masbury summit with No 34040, *Crewkerne*, on a Bournemouth bound express in August 1962. At the end of the Summer 1962 timetable all through services were diverted away from the S&D and this caused No 75027 to move on once more.

M. J. Esau

88
By early 1963 No 75027, made redundant on the S&D, had moved to the Cambrian lines and was based for a short period at Machynlleth before moving to Croes Newydd depot near Wrexham. It is seen entering Bala Junction, on the Barmouth–Ruabon line, with the 2.35 p.m. Barmouth–Chester train on 23rd July, 1963.

B. Stephenson

89
On the same route No 75027 enters Drws-y-nant with the 12.45 p.m. Pwllheli—Wrexham on 27th July, 1963.

M. Hale

90
A few weeks later the Cl. 4 pulls away from Llangollen with a Chester-bound train. This highly scenic route is now completely closed to traffic. Llangollen Station however has been taken over by a preservation group.

R. E. Toop

91

In the mid-sixties No 75027 again found itself on the move and this time it was sent north to Lancashire. By 1968 it was based at Carnforth and found employment on a variety of turns including banking on Grayrigg Incline between Oxenholme and Tebay. It also worked stone trains from Swinden quarry, on the Grassington branch near Skipton, and is seen here at the Works in May 1968.

Dr. L. A. Nixon

92

On 21st June, 1968, No 75027 leaves Embsay Tunnel in charge of the daily ballast train from Swinden quarry. These trains were among the last regular workings for the class.

Dr. L. A. Nixon

93
With just one week to go before the end of BR steam No 75027, piloted by sister engine No 75019, waits to leave Carnforth with an enthusiasts special organised by the Severn Valley Railway Society.

N. E. Preedy

94
Later the same day the pair of Cl. 4s make a fine sight between Bentham and Clapham on the Carnforth–Skipton line.

J. G. Mallinson

95
Following its arrival at Sheffield Park in January 1969, ▷ No 75027 worked for a short period in its shabby BR paintwork before being withdrawn from traffic for a complete re-paint. This work was completed in August 1971, and here a gleaming No 75027 is seen starting the stiff climb from Holywell to Horsted Keynes a few weeks after re-entering traffic.

M. J. Esau

SE & CR 0-6-0T No 178

96
No 178 was the last of the Bluebell Railway's three P class tanks to arrive on the line; it was acquired in October 1969. When new No 178 was set to work on the Reading–Ash–Aldershot auto-train services, but later moved to the Otford–Sevenoaks line. It was working this service when photographed at Sevenoaks Station in 1922.

Courtesy, National Railway Museum, York

97
Another shot of No 178 at Sevenoaks, this time on 27th September, 1923.

LCGB Ken Nunn collection

98

In early BR days the locomotive ran for a time as No s1178; this picture was taken at Brighton.

Lens of Sutton

99

On 1st November, 1957, No 31178 was acting as shed pilot at Stewarts Lane.

P. H. Groom

100

In June 1958, No 31178 was sold out of service to Messrs. Bowaters of Ridham Dock, near Sitting-bourne, who had previously hired the same locomotive for a short period in 1953. In Bowaters service, it was named *Pioneer II*. Here it stands at their factory complex at Ridham Dock on 30th July, 1968.

P. H. Groom

101
Stamford poses at South Pilton on 21st May, 1966.
T. J. Edgington

AVONSIDE 0-6-0ST No 24 'STAMFORD'

On a line dominated by ex-BR motive power the industrial locomotives on the Bluebell might be considered out of place. Nonetheless, No 24 *Stamford* is an attractive engine and when restored is likely to be a useful locomotive for pilot or shunting duties.

No 24 was built by Messrs. Avonside of Bristol for the Staveley Coal and Iron Company Limited, for use at their ironstone quarry at Pilton in Rutland. The locomotive spent almost its entire life working there hauling wagons between the quarry and the main line exchange sidings. Later the quarry was worked by Stewarts & Lloyds and in the late nineteen-sixties they rationalised their ironstone quarrying activities with the result that *Stamford*, together with many other similar locomotives, became surplus. During early 1969 a party of Bluebell members inspected the locomotives and selected No 24 as being the best available. *Stamford* was delivered to Sheffield Park in October 1969, and is currently dismantled for heavy overhaul.

102
Another picture of *Stamford*, this time on rail-tour duty at the quarry, thought to have been taken on the same day.

Bluebell Archives

103

Built in February 1902, No 592 is a fine example of an inside cylinder 0-6-0 goods engine, a type that was once commonplace throughout the British railway system. This locomotive has a particular claim to fame as the only remaining Longhedge-built locomotive. The type was developed from the LCDR B class and was so successful that by 1908 a total of 109 machines had been produced and the type later became the largest class numerically on the Southern Railway. Under Southern management, No 592 became No 1592, and it is seen posing in New Cross Gate yard on 5th July, 1932.

H. N. James

104

The C class was frequently employed on comparatively long distance excursion trains, and in this delightful picture No 1592 approaches Chelsfield in charge of a rake of LCDR six-wheel coaches. This photograph was taken in 1935 just before electrification.

N. E. Preedy collection

SE & CR 0-6-0 No 592

105
No 31592 in early BR livery.
W. M. J. Jackson

106
A begrimed No 31592 shunts
Ramsgate carriage shed on 16th
August, 1958.

D. J. Wigley

107
On a Cuckoo line goods near
Heathfield Tunnel in March 1961.
S. C. Nash

108

With 01 0-6-0 No 31065 as pilot No 31592 waits to leave Cranbrook with the LCGB 'South Eastern Limited' rail tour on 11th June, 1961. This was the last train along the Hawkhurst branch which closed to passengers the following day.

T. Stephens

109

Later the same day the pair were photographed near Pattenden Siding on the return stage of their melancholy journey back to Paddock Wood.

D. Cross

110

Nos 31271 and 31592 (nearest camera) take water at New Romney during snow clearing duty on New Year's Day 1963.

G. Barlow

111

Later in 1963, No 31592, together with two sister engines, was transferred to departmental stock as an Ashford Works shunter. It was re-numbered DS 239 and remained in the Work's employ until replaced by USA class 0-6-0Ts early in 1965. On withdrawal from service the locomotive was purchased by the Wainwright C Class Preservation Society, which had been raising funds for this purpose since 1962. The engine is seen here undergoing restoration in the Works area on 28th May, 1967.

J. G. Mallinson

112
Prior to removal to the Bluebell the engine spent a few years accommodated at the privately run Ashford Steam Centre where considerable restoration work was undertaken.

T. Stephens

113
The 'C' class in silhouette, pictured on the Bluebell in 1977.

M. J. Esau

SR 4-6-2 No 21C123
'BLACKMORE VALE'

114

Blackmore Vale is without doubt one of the most impressive preserved locomotives in the country. It was built to the design of O.V.S. Bulleid, the Southern Railway's C.M.E. from 1937 to 1949. Bulleid's innovatory ideas caused considerable controversy, but in many respects his Lightweight Pacifics were outstanding in design and performance. No 21C123 was built at Brighton Works in February 1946 and was initially based at Ramsgate. This picture shows the locomotive—then un-named—at Ashford (Kent) working a down Folkestone express just five months after its construction.

D. W. Winkworth

115

Sunlight and shadow at Exeter St. Davids. No 34023 simmers gently after bringing in a train from Ilfracombe on 6th December, 1962.

J. R. Besley

116
Waterloo Station was the last London terminus to see regular steam operation, and became a 'Mecca' for many enthusiasts. *Blackmore Vale* awaits departure with a Bournemouth train in July 1964.

G. R. Siviour

117
Blackmore Vale approaches Honiton Tunnel with the 10.15 (SO) Waterloo–Ilfracombe on 22nd August, 1964.

Ivo Peters

118

When engineering work in connection with the Bournemouth line electrification scheme was in progress in the mid-sixties, many diversions took place particularly at weekends. On Sunday, 13th March, 1966 No 34023 worked the 10.30 a.m. Waterloo–Bournemouth which was routed via the Portsmouth Direct line. Here the train is seen climbing the 1 in 80 gradient between Witley and Haslemere.

J. S. Everitt

119

On 12th September, 1966, *Blackmore Vale* speeds through the New Forest near Beaulieu Road in charge of the up 'Bournemouth Belle'.

D. T. Cobbe

120

Night Duty! The following day No 34023 found itself working the 9.25 p.m. Waterloo Bournemouth. It is pictured prior to departure from Waterloo.

Dr. L. A. Nixon

121
The low evening sun glints against the side of *Blackmore Vale*.

P. J. Fowler

122
After withdrawal from BR service, No 34023 found a temporary home on the Longmoor Military Railway in Hampshire where it is pictured in light steam. *Clan Line* is just visible in the shed.

Bulleid Society Archives

123

A scheme to establish a preservation centre at Longmoor met with opposition from local residents and had to be abandoned. Consequently a new home had to be found for No 34023 and the locomotive's owners, the Bulleid Society, decided on a move to the Bluebell. The locomotive arrived during September 1971 and the long, arduous task of restoration commenced. Four years later, on 2nd November, 1975, a steam test was arranged and this proved so successful that a trip was made up the line. Here the locomotive is depicted making a dramatic ascent of Freshfield bank.

M. J. Esau

124

On 15th May, 1976, *Blackmore Vale*, now re-numbered 21C123, made its official return to service—the culmination of nearly ten years voluntary effort by the Bulleid Society. Here an immaculate No 21C123 eases out of Sheffield Park with the last train of the day.

B. E. Morrison

125
No 30064 began its career as U.S. Army Transportation Corps No 1959 at the Vulcan Ironworks, Wilkes Barre, Pennsylvania, USA, in 1942. On arrival in Britain No 1959 was sent to the Melbourne branch of the LMS in Leicestershire for slight modifications to be carried out prior to entering service as part of the Allies war back-up. It is doubtful if the locomotive ever turned a wheel in war service, and by 1947 it was gathered with others of its type at Newbury Racecourse as war surplus material. It is pictured there on 15th March, 1947.

A. W. Croughton

126
Six months later, No 1959 had been purchased by the SR for use in Southampton Docks; it cost just £2,500, becoming No 64 in their series. Before entering service it received a thorough overhaul at Eastleigh, and various modifications were made, particularly to the cab window arrangement. It is seen here at Southampton Docks shed on 21st September, 1947.

H. C. Casserley

SR 0-6-0T No 30064

127
The 'Yankee Tank', as the USAs were known colloquially, stands in the Southampton Docks area on 15th August, 1951. The number, s64, and British Railways legend on the tanksides, are almost hidden by a thick coating of grime.

A. N. H. Glover

128
Five years later No 30064, as it became, was still engaged on dock shunting and is seen at the Old Docks in October 1956. In 1962 new diesel shunters ousted the USAs and No 30064 was redeployed on pilot duties at Eastleigh.

P. S. Leavens

129
In 1963 the SR had the happy idea of repainting some USAs in malachite green livery much to the chagrin of the Design Panel at Marylebone. No 30064 was one of those treated and was out-shopped in February 1964. It was temporarily transferred to Guildford soon afterwards, for duty as the shed pilot, and is seen there on 29th May, 1964.

J. R. Besley

130

Later that year the USAs attracted the attention of rail-tour sponsors and No 30064 was employed on the Woking–Horsham leg of a joint RCTS/LCGB tour on 18th October, 1964. It is pictured near Bramley and Wonersh, and appears to be coping admirably with its well laden eight coach load.

B. Stephenson

131

On 20th March, 1966, No 30064 was again on rail tour duty. Here it partners sister engine No 30073 along the Fawley branch. Later that year No 30064 achieved fame when it was transferred to Meldon Quarry, near Okehampton, and subsequently became the Western Region's last active steam locomotive.

D. Trevor Rowe

132

A portrait of No 30064 in its final BR condition; this picture was taken at Eastleigh in May, 1967, two months before the end of Southern Region steam.

J. G. Dewing

133
When steam finally came to an end, No 30064 joined the
long lines of redundant locomotives on Salisbury dump
where it is pictured waiting disposal on 7th October, 1967.
The locomotive immediately behind is No 30072 which is
now preserved on the Keighley and Worth Valley line in
Yorkshire.

E. Wilmshurst

134
Following purchase from BR by the Southern Locomotive
Preservation Company early in 1968, No 30064 was stabled at
Droxford, and later at Liss, before coming to the Bluebell in
October 1971. Since then it has rarely been out of service,
but is not regularly used during the Summer owing to its tendency
to ignite crops in the fields adjoining the line! Here it is seen
approaching Freshfield in charge of a Sheffield Park—Horsted
Keynes train on 26th May, 1975.

B. E. Morrison

135
No 96 was built in 1893 by the LSWR for dock shunting. The LSWR bought out the Southampton Dock Company in 1891 and required a standard shunter to replace the motley collection of locomotives inherited from the Dock Company. Ten locomotives were built initially to Order No B4 (hence the class designation) and further machines were constructed later giving a total of 25 in the class; *Normandy* was one of this later batch. It spent the first fifty years of its life on shunting duties in the docks and is seen there on 31st July, 1930. The cut-away cab, fitted for improved visibility, will be noted. The apparatus behind the dome is a linseed filtrator which was fitted to this type of locomotive from 1927 onwards to prevent the formation of boiler scale; they were later removed.

LCGB Ken Nunn collection

L & SWR 0-4-0T No 96 'NORMANDY'

136
In 1947, the Southern Railway purchased a batch of war surplus American switchers (class USA) to replace the B4s, as many of the latters' boilers required renewal. Some B4s were, however, retained for special duties requiring short wheelbase engines, and No 96 was one of these. It is pictured shunting at Winchester on 6th August, 1949.

Bulleid Society Archives

137
In 1959, No 30096, as it was known under the BR numbering scheme, received a thorough overhaul at Eastleigh Works and is seen here in the erecting shop on 2nd June, 1959.

J. Oatway

138
Southampton Docks depot on 8th June, 1961; the locomotive lurking in the background is an E2 0-6-0T.

J. C. Haydon

139
No 30096 shunts Winchester goods yard on 24th June, 1961. This was the last duty for the class which was not displaced here until the Autumn of 1963.

J. C. Haydon

140
On 9th March, 1963, No. 30096 worked a rail tour in the Solent area. Here it is seen posing alongside the Ocean Liner Terminal in Southampton Docks.

J. C. Haydon

141

On withdrawal from BR service in October, 1963, No 30096 was acquired by Corralls, the fuel merchants, to shunt their private wharf at Southampton. Renamed *Corrall Queen*, it was retained on these duties until December 1972 when it was purchased by the Bulleid Society and brought to Sheffield Park. Here the engine is pictured on the wharf dwarfed by a massive travelling crane. The departure of No 30096 ended the class's eighty year association with the Southampton area.

J. G. Mallinson

142
No 263 was one of the ubiquitous H class 0-4-4Ts built between 1904 and 1909: No 263 was constructed at Ashford Works in May 1905. It spent the first twenty years of its life at Slades Green depot employed on suburban duties in South East London. When these routes were electrified it was transferred to the country, and earned its keep on Kentish branch lines. It is seen at Dover shed on 25th June, 1939.

H. C. Casserley

SE & CR 0-4-4T No 263

143
In final SR condition No 1263 poses at Stewarts Lane depot on 10th May, 1949.

J. Kent

144
A Hawkhurst branch train leaves Paddock Wood behind No 31263 in August 1960.

J. Kent

◁ 145
Shortly before electrification No
31263 leaves Yalding on a Maidstone
West—Tonbridge working in
September 1960. The coach
immediately behind the engine is
No S971S which is also now on the
Bluebell line.

D. Cross

146
On 25th February, 1962, No 31263 was employed on
an LCGB rail tour which marked the virtual end of
steam traction in Kent. Here No 31263 paired with
'C' class No 31690, awaits departure from Ashford.

L. W. Rowe

147
The pair pull away from Lydd en route to
New Romney.

G. R. Siviour

148
With the spread of modernisation the work for push-pull engines was steadily reduced and No 31263 was drafted to the Three Bridges—East Grinstead line on the Central Division. It is pictured at Rowfant propelling an East Grinstead-bound "push-pull" train on a Summer day in 1962. On the right K class No 32349 waits patiently with an engineers' train for Three Bridges.

G. R. Siviour

149
No 31263 makes a fine sight as it sets off from Grange Road on 6th April, 1963.

D. T. Cobbe

150
On an appropriately gloomy day No 31263
stands at Three Bridges on 4th January, 1964.
The engine had just completed its last duty in
BR service, on a train from East Grinstead, and
waits to berth its stock before retiring to the
shed for the last time. One of the replacement
DEMUs waits in the bay.

J. G. Mallinson

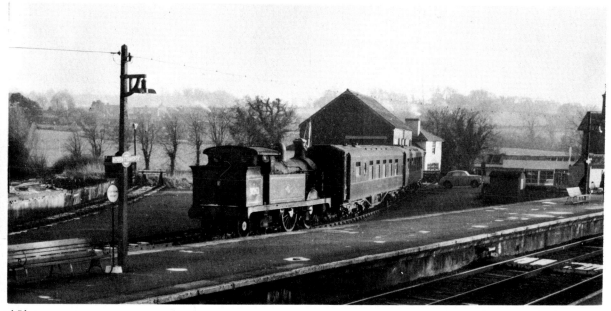

151
After a period of storage at Three Bridges
No 31263 was purchased by the H Class Trust,
and moved to Robertsbridge by freight train
on 11th November, 1964. Here it joined two
Pullman cars which were destined for eventual
use on the Kent and East Sussex line.

D. T. Cobbe

152
The engine later moved to the privately owned
Ashford Steam Centre and was photographed
there during an "Open Day" in October 1971.
At this time the locomotive had been partially
restored.

D. T. Cobbe

153
Study in steam and cloud!
No 263 arrived on the Bluebell
early in 1976 and has since
been a regular performer;
this picture was taken in
April 1978.

M. J. Esau

154
No 263 and C class No 592
make a superb combination
as they leave Sheffield Park
in charge of the 3.55 p.m. to
Horsted Keynes on Vintage
Transport Sunday, 10th
September, 1978.

S. C. Nash

SR 0-6-0 No C1

156

For a long period No 33001, as C1 became under the BR re-numbering scheme, was based at Guildford shed. It is seen here on 17th May, 1959, heading the 8.47 a.m. Reading South–Brighton excursion between Slinfold and Christ's Hospital on the now closed Guildford–Horsham line.

J. Scrace

155

No 33001 was constructed as SR No C1 in 1942 at Brighton Works. It was the first of a class of 40 locomotives built under wartime conditions to austerity standards. As a result of the wartime phonetic code ('C for Charlie'), the engines were commonly known as 'Charlies'. In service they were noted for rugged reliability, and proved to be versatile machines, being equally at home on both passenger and freight work. In this view No C1 stands at Redhill on a ballast working on 16th August, 1947.

LCGB Ken Nunn collection

157
This portrait was taken on 16th August, 1960, at Tonbridge depot. Note the scorched smokebox door.

J. Scrace

158
No 33001 heads an evening Southampton-bound van train near Wimbledon in August 1961. By this time the engine had been re-allocated to Feltham.

P. F. Winding

159
A Guildford–Horsham train with 33001 in charge leaves Bramley and
Wonersh in 1961.

M. J. Esau

160
At Nine Elms during the Winter of 1962.

M. J. Esau

161
No 33001 heads the 2.00 p.m. Readir
—Feltham goods near Winnersh
Halt on 1st March, 1963.

J. S. Everitt

162
Following withdrawal in May 1964, No 33001 was set aside for official preservation by
BR, and was put into store. Before being placed on loan to the Bluebell Railway in 1977
the locomotive had been kept in the former Pullman Car works at Preston Park, just
outside Brighton. During its stay there it was occasionally displayed on "Open Days"
held at Brighton Station, and is seen at one of these events in 1974.

J. G. Mallinson

163

No 1618 was constructed at Brighton Works in 1928, to the U class design of R.E.L. Maunsell. Commonly described as a 'Southern Mogul', the type was familiar throughout the Southern Railway system for almost forty years. In this picture No 1618 stands at Redhill prior to departure with a through Ramsgate–Birkenhead train on 31st March, 1939.

H. M. Madgwick collection

SR 2-6-0 No 1618

164

For many years the 'Moguls' worked the attractive Midland & South Western Junction line. Here No 31618 pauses at Swindon Town while working a Cheltenham–Southampton train on 28th September, 1957. In time honoured fashion the driver 'oils round' while the engine is watered.

D. W. Winkworth

165 and 166
On 29th August, 1959, No 31618 spent the day working between Salisbury and Portsmouth. It approaches Fareham in charge of a morning train from Salisbury . . . . . . and later leaves Fareham in the evening sun with its return working. Note the superb rake of Maunsell coaches.

J. Kent

167
The 'Mogul' was fresh out of shops when this picture was taken of her leaving Bournemouth Central with the 9.40 a.m. Brighton–Bournemouth West train on 26th August, 1961.

R. A. Panting

168
No 31618 coasts towards Basingstoke with a down Waterloo–Basingstoke semi-fast in September 1961.

P. S. Leavens

169

During the declining years of SR steam the U class was largely congregated at Guildford shed. On New Years Day 1964, No 31618 accelerates away from Guildford on a Reading train, passing S15 No 30842 (not No 30847 alas!) waiting in the yard.

D. E. Esau

170

On 20th June, 1964, No 31618 left Fratton, where it had been sent following condemnation, for Barry scrap-yard in South Wales. It is pictured at Barry on 27th August, 1968, still looking presentable after four years' exposure to the elements.

D. Idle

171

No 31618 was one of the first locomotives to leave Barry and was purchased in reasonably complete condition. It found a temporary home at a paper mill at Aylesford, Kent, but later it was moved to the Kent and East Sussex Railway and kept at Tenterden where restoration was completed. However the locomotive was too heavy for the K & ESR and was thus unable to venture beyond the confines of Tenterden yard. In 1977 a move to the Bluebell was arranged and on 17th May, the locomotive arrived at Sheffield Park following rapid negotiations with the haulage contractors who were moving the Q1 from Preston Park at the same time. Since its arrival on Bluebell metals No 1618 has been a revelation, being consistently free-steaming, reliable and economical. Here it illustrates its versatility by powering a Bluebell freight working on 13th May, 1979.

J. G. Mallinson

172
On 5th July, 1962, BR operated an excursion from Somersham (Huntingdonshire) to Gloucester. The return train is seen here approaching Andoversford behind No 92240.

J. Dagley-Morris

On the Bluebell it has been found that the larger steam locomotives have a special appeal. If this is true there can be no doubt that No 92240, with the locomotive alone weighing nearly ninety tons, is destined to become a firm favourite. The 9Fs were designed at Brighton and became one of the most successful BR Standard types. They were amazingly versatile and despite being built primarily for freight work have shown a fair turn of speed on passenger duties. On release from Crewe Works in 1958, No 92240 went to the Western Region and spent most of its seven years active life based at Southall Depot.

BR STANDARD 2-10-0 No 92240

173
No 92240 appears to have been rarely photographed during its BR career, as gleaming WR 'Kings' and 'Castles' were no doubt considered a worthier subject for the camera. However the 9F was caught on film on 19th October, 1963, passing West Drayton with an up freight.

B. Stephenson

174
The Class 9F drifts through High Wycombe with a north-bound bulk cement train on 10th May, 1965.

R. Potter

175
No 92240 stands at Oxford depot on 17th July, 1965. It was withdrawn later that year after a mere seven years in BR service. Not surprisingly, an examination by Bluebell Railway engineers found the locomotive to be in excellent basic condition and well worthy of preservation. It arrived at Sheffield Park in October 1978 and phenomenal progress has since been made towards restoration to full working order.

N. Preedy

SR 0-6-0 No 541

176
Like so many locomotives based at Sheffield Park, No 30541 is the only one of its type surviving. No 541 was one of a batch of 20 ordered at a cost of £7,200 each in March 1936 to Maunsell's specification; it was not until January 1939, however – after O.V.S. Bulleid had taken over as CME – that No 541 emerged from Eastleigh Works. In December 1946 No 541, in common with all its sisters, was fitted a Lemaître multiple jet blastpipe and large diameter chimney. In this picture though, taken at Norwood Junction depot on 28th July, 1945, No 541 is still in original condition.

H. C. Casserley

177
During the late nineteen-forties, No 30541 worked on the Oxted line for a period. Here it powers a down train near Sanderstead in 1949.

R. W. Beaton

178

For some years, No 30541 was based at Bournemouth and was a familiar sight on the lines in Hampshire. Here it is seen working the 1.28 p.m. Lymington Pier–Waterloo along the Lymington branch near Ampress Halt on 5th September, 1959.

S. C. Nash

179

No 30541 wheels a heavy freight through Southampton Central in June 1960. The attractive clock tower at this station, for so long a familiar landmark to Bournemouth line travellers, disappeared when the station was re-built in the late nineteen-sixties.

D. Cross

180
No 30541 ambles into Wareham with a short pick-up
freight on 26th June, 1961.

R. White

181
Shunting empty coaching stock at Brockenhurst in August
1961.

D. T. Cobbe

182
No 30541 sits out in the sun at Guildford shed on 13th July, 1964.

R. Blencowe collection

183
No 30541 was withdrawn in November 1964 and eventually found its way to Barry scrapyard. In 1973 a fund was started to rescue the locomotive; it was subsequently bought for preservation and removed to a site at Ashchurch, Gloucestershire. In October 1978 this distinctive machine was transferred to the Bluebell line where restoration to full working order is being undertaken. The engine is seen here making a "safe landing" from the road transporter at Sheffield Park.

P. D. Nicholson

184
No 80100 was built at Brighton in 1955 and was initially allocated to Plaistow depot, East London, for duty on the ex-LT & SR lines from Fenchurch Street. On 24th August, 1957, it was photographed approaching Barking powering an eastbound train; at this time the LT & SR section was in the throes of reconstruction prior to electrification.

A. A. Jackson

BR STANDARD 2-6-4T No 80100

As the Bluebell's collection of locomotives gradually expanded in the nineteen-seventies it became clear that there was one major omission— the line was still not able to boast a Standard Class 4 tank. The class's strong links with Brighton Works and the closing years of the Bluebell line under BR, made the acquisition of one of these machines essential to complete the Railway's locomotive collection, and in 1976 a fund was established to purchase a member of the class from Barry scrapyard. By this time No 80100, the locomotive selected, had lain there for eleven years, but was none-theless declared capable of restoration.

185
No 80100 stands in Fenchurch Street Station in October 1961.

J. G. Mallinson

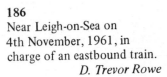

186
Near Leigh-on-Sea on 4th November, 1961, in charge of an eastbound train.
D. Trevor Rowe

187

With the electrification of the LT & SR lines in
June 1962, No 80100 was out of work, and
found itself shipped over to the other side of
London to Old Oak Common depot where,
along with half a dozen other refugees from
Plaistow, it spent a few weeks in store while
the Western Region decided what to do with it.
Eventually No 80100 was allocated to
Shrewsbury and found employment on local
passenger work, including trips on the Severn
Valley line, part of which is now preserved. On
27th October, 1962, No 80100 was photographed
entering Ironbridge and Broseley with the
11.35 Bridgnorth–Shrewsbury.

J. Spencer Gilks

188

Minus its front number plate No 80100 looks
rather anonymous as it leaves a tunnel on the
Grange Court Junction–Hereford line in 1963.

J. Dagley-Morris

189
No 80100 leaves the delightfully situated station of
Mitcheldean Road, also on the Grange Court—Hereford
line. The train is the 12.15 p.m. Gloucester—Hereford,
and the date is 26th September, 1964.

D. Capewell

190
In October, 1978, No 80100
was resurrected from Barry
scrapyard, and in this picture
is seen near Caterham in the
course of delivery to the
Bluebell line.

J. G. Mallinson

191

The Maunsell S15s were a development of the Urie S15s of 1920–21, and were constructed in two batches. The first batch was built in 1927–28, but No 847, part of the second series, was not constructed until December 1936 and was the very last 4-6-0 locomotive to be built for the Southern Railway. It is recorded that each engine of this later series cost £14,145 to construct. No 847 was barely a year old when this picture of her was taken at Raynes Park in 1937 working an up Bournemouth Line train.

Bournemouth Railway Club Kelland collection

SR 4-6-0 No 847

192
Ten years later No 847 is seen near Templecombe in final SR condition.

Locomotive and General Railway Photographs

193
Near Winchfield with an up Salisbury line stopping train on 16th May, 1953.

K. W. Wightman

194
A shaft of sunlight breaks through an otherwise overcast sky as No 30847 eases out of the yard at Axminster with an eastbound freight. This picture was taken on 10th July, 1956.

R. C. Riley

195
No 30847 waits to leave Exeter Central with the 3.20 p.m. local train to Templecombe on 3rd July, 1958.

J. Scrace

196
Rounding the curve at Shalford with a Reading—Redhill freight on 12th December, 1961. By this time the S15 had moved from the South Western Section to Redhill on the Central Section. As a result of this the locomotive lost its 5,000 gallon eight-wheel tender in exchange for a 4,000 gallon six-wheel tender from a withdrawn 'King Arthur' class 4-6-0. The change enabled the locomotive to be accommodated on the smaller turntables used on the Central Section.

D. Fereday Glenn

197

At Reading shed on 31st May, 1963.

P. H. Groom

198

No 30847 was finally withdrawn by BR in January 1964, and despatched to Barry on the 18th June of that year. It travelled in a convoy from Feltham shed with three other S15 s; coincidentally two of the other locomotives, Nos 30506 and 30841, are also now preserved. Here we see No 30847 waiting to leave Barry for Sheffield Park on 8th October, 1978.

M. J. Allen

199

No 73082 was one of a class of 172 Standard 5MT 4-6-0s turned out over a six-year period from 1951 to 1957. No 73082 was built at Derby Works and entered traffic in June 1955, based at Stewarts Lane depot. The locomotive spent the first few years of its brief career working from Victoria on Kent Coast expresses, and boat trains to Dover or Folkestone. In this view it is seen threading its way through Herne Hill with the 8.35 a.m. Victoria–Ramsgate in June 1956.

D. W. Winkworth

BR STANDARD 4-6-0 No 73082 'CAMELOT'

200

In 1958 during its allocation to the South Eastern Division, with electrification reaching an advanced stage by this time, No 73082 was photographed on 13th September, 1958, heading a down passenger working near Whitstable.

D. W. Winkworth

201
No 73082 passes St. Mary Cray Junction on 16th
May, 1959, in charge of a down Kent Coast Express.
Later that year the SR decided to revive the names of
withdrawn Urie 'King Arthur' class locomotives and
No 73082 was allotted the name *Camelot*.

R. C. Riley

202
Later the same year, No 73082 left Stewarts Lane
and was moved to another London shed, Nine Elms,
from where it powered services from Waterloo. It
was photographed in September 1963 approaching
Basingstoke at speed on a down Bournemouth line
train.

G. R. Siviour

203
Camelot passes Vauxhall with the 11.30 a.m. Waterloo—
Bournemouth on 25th September, 1964.

B. Stephenson

204
Camelot was withdrawn on 19th June, 1966,
and was dumped at Eastleigh prior to
removal to Barry scrapyard where it arrived
five months later. This picture, taken just
before the engine was moved to the Bluebell,
tells its own story. No 73082 arrived on
Bluebell metals on 27th October, 1979,
after five years of fund raising by the
Camelot Locomotive Society.

M. J. Allen

205

Sir Archibald Sinclair is slightly younger than *Blackmore Vale*, being out-shopped from Brighton Works as 21C159 in April 1947. When new the engine was allocated to Nine Elms.

Courtesy, National Railway Museum, York

SR 4-6-2 No 34059 'SIR ARCHIBALD SINCLAIR'

206

Sir Archibald Sinclair was officially named at a ceremony held at Waterloo Station in February 1948.

Courtesy, National Railway Museum, York

207
During May 1949, after re-numbering to 34059, the engine was based at Stratford shed in East London for trials on the Great Eastern line out of Liverpool Street. On 18th May No 34059 powered the 'Norfolkman' from London and is pictured awaiting departure from Norwich Thorpe with the return train.

LCGB Ken Nunn collection

208
Later the same afternoon *Sir Archibald Sinclair* was photographed passing Church Lane crossing, near Ingatestone, at speed. Note the smoke drifting down to obstruct the view from the foot-plate; this was always a problem with Bulleid Pacifics which was only partially cured by rebuilding.

Bulleid Society Archives

209
Following its return to the SR *Sir Archibald Sinclair* was photographed south of Dorchester on 26th September, 1949, hauling a down Weymouth train.

D. K. Jones collection

210
For a brief period following re-building No 34059 operated on the South Eastern Division; it is seen here entering Waterloo Eastern with the 9.10 a.m. Charing Cross–Ramsgate on 18th May, 1961.

J. Scrace

211
With just a few weeks to go before the introduction of full electric working *Sir Archibald Sinclair* rushes through Sandling with a down express.

D. T. Cobbe

212
On a damp evening *Sir Archibald Sinclair* waits for the road
away from Exeter Central on Whit Monday 3rd June, 1963,
with the 5.30 p.m. express to Waterloo.

J. R. Besley

213
No 34059 approaches Milborne Port with a train of milk
tanks in tow on 18th May, 1964.

M. Mensing

214
Sir Archibald Sinclair leaves Seaton Junction in charge of the 11.48 a.m. Plymouth–Waterloo on 18th July, 1964.

Ivo Peters

215
Hauling less than her own weight No 34059 dashes through Vauxhall at the head of a Basingstoke semi-fast in August 1964.

R. L. Sewell

216
Sunlight and shadow at Nine Elms depot on
23rd October, 1965.

P. S. Leavens

217
In late 1978 a group of members formed the
Bluebell Railway Battle of Britain Locomotive Group with the aim of saving
No 34059 for eventual use on the line. In
less than a year they raised £7,250 needed to
purchase the locomotive from Messrs.
Woodham's yard at Barry, and the machine
is depicted here arriving at Sheffield Park on
27th October, 1979. The locomotive is
carrying display boards advertising Gateway
Natural Tracing Paper, which is manufactured
by Wiggins Teape Ltd., who generously
donated £3,250 to cover the cost of the road
movement.

M. J. Allen

Bluebell Steam...

9F driving wheels.

M. J. Esau

263's Ramsbottom safety valves.

M. J. Esau

Adams Radial front end.

M. J. Esau

Clack valves on USA tank 30064.

M. J. Allen

Front bufferbeam number on *Bluebell*.
M. J. Esau

Blackmore Vale's nameplate and Dorset coat-of-arms.

M. J. Esau

Live steam injectors on *Blackmore Vale*.
M. J. Allen

Birch Grove's Westinghouse brake pump.
M. J. Esau

Walschaerts valve gear on 'Standard 4' 75027.
M. J. Esau

...in Close-up

The last train of the day leaves Sheffield Park on a Winter afternoon. But for the dedicated support of the Bluebell Railway Preservation Society the Railway would not survive today. The Society is composed of ordinary people from many walks of life. If you consider that the continued preservation of the Bluebell is worthwhile, why not join the BRPS?

Write to :　The Membership Secretary
　　　　　The Bluebell Railway Preservation Society
　　　　　Sheffield Park Station
　　　　　Nr. UCKFIELD
　　　　　East Sussex
　　　　　TN22 3QL